# Topical Conference: Material Interfaces as Energy Solutions



Held at the 2023 AIChE Annual Meeting

Orlando, Florida, USA
5-10 November 2023

ISBN: 978-1-7138-9316-5

**Printed from e-media with permission by:**

Curran Associates, Inc.
57 Morehouse Lane
Red Hook, NY 12571

**Some format issues inherent in the e-media version may also appear in this print version.**



AIChE
120 Wall Street, FL 23
New York, NY 10005-4020

Phone: (800) 242-4363
Fax: (203) 775-5177

www.aiche.org

**Additional copies of this publication are available from:**

Curran Associates, Inc.
57 Morehouse Lane
Red Hook, NY 12571 USA
Phone: 845-758-0400
Fax: 845-758-2633
Email: curran@proceedings.com
Web: www.proceedings.com

# TABLE OF CONTENTS

## PLASMA CATALYSIS

335a Elucidating Non-Thermal Plasma-Activated Carbon-Nitrogen Coupling on a Catalytic Surface from Methane and Nitrogen ..... 1
*Garam Lee, David Go, Casey O'Brien*

335b Pathways and Timescales for $N_2$ Conversion By Reductive or Oxidative Routes in Plasma Catalysis ..... 2
*Brian Bayer, Peter Bruggeman, Aditya Bhan*

335c Elucidating Plasma-Catalyst Contributions for the Direct Conversion of Methane to Aromatics ..... 3
*Gerardo Rivera-Castro, Ibukunoluwa Akintola, David Go, Jason Hicks*

335d Machine Learning Boosted Design of Plasma-Enhanced Catalytic Ammonia Cracking ..... 4
*Saleh Ahmat Ibrahim, Hsi-Wu Wong, Juan Pablo Trelles, Fanglin Che*

335e Selective Propylene Epoxidation from Plasma-Liquid Interactions Using Water As the Oxygen Source ..... 5
*Dongho Lee, Han-Ting Chen, Suljo Linic*

335f Decoupling Plasma-Catalytic Interactions through Plasma-Temperature Programmed $N_2$ Reduction and Oxidation Reactions ..... 6
*Hope Otor, Patrick Barboun, Hanyu Ma, Anshuman Goswami, William Schneider, Jason Hicks*

Mn-Ferrite As a Highly Active Catalyst for Toluene Degradation in Post-Plasma Catalysis ..... 7
*Khurram Shahzad Ayub, Waqas Qamar Zaman, Zain Abbas*

335g Evaluate the Feasibility of Employing Cactus Mucilage to Improve the Robustness of the Reactive Coating Photocatalyst for the Removal of Chosen Bacteria in Wastewater ..... 8
*Khuloud Alrashdi, Norma Alcantar*

## MODEL BASED DESIGN AND OPTIMIZATION FOR HYBRID MATERIALS AND INTERFACES

257a Physics and Data-Informed Formulation Design and Development for Chemical Processes ..... 9
*Shikha Mallick, Francesca Boioli, Ravindra Aglave, Panos Petris, Peter Mas*

257b Data-Driven Optimization of Perovskite Solar Cells ..... 11
*Donghyun Oh, Sanggyun Kim, Carlo Andrea Riccardo Perini, Juan-Pablo Correa-Baena, Nikolaos Sahinidis*

257c Controlling Structure-Property Relationships in Organic Semiconductors Using Tunable Highly-Ordered 2D Perovskite Interfaces ..... 12
*Meagan Phister, Ashley Conley, Rachel Dziatko, Karl Westendorff, Anna Sviripa, Stephanie Guthrie, Kevin Stone, Arthur Bragg, Christopher Paolucci, Joshua Choi, Gaurav Giri*

257d Benchmark Simulation of Ag Nanoparticles on Supported Graphene Substrates ..... 13
*Tristan Maxson, Tibor Szilvasi*

257e Effects of the Polymer/MOF Interface in Mixed-Matrix Membranes on Small Molecule Transport ........ 14
*Moein Z.D. Hajian, Hyunhee Lee, Zachary P. Smith, Joshua Howe*

257f Utilizing Tunable Surfaces for Performance and Stability Enhancement of Nitramine Energetic Materials ........ 15
*Natalie Smith, Meagan Phister, Gaurav Giri*

257g Photochemical Bubble Generation on Polymer-Water Interface. ........ 16
*Pranaya Ghate, Christopher J. Bardeen, Rabih O. Al-Kaysi*

## MATERIAL INTERFACES FOR ENERGY APPLICATIONS

212a Investigating the Coupled Effect of IFT Reduction and Wettability Alteration in Enhanced Oil Recovery from Carbonate Reservoirs ........ 17
*Muhammad Shahzad Kamal, Deng Xiao, Shirish Patil, Syed Hussain*

212c Chemical Oscillation at the Grain Boundary of Aliovalently-Doped Solid-Oxide Electrolytes ........ 18
*Bharathi Bandi, Abhijit Chatterjee*

212e Impact of Cathode/Solvent Interfaces on the Electrochemical Performance of Na-$O_2$ Batteries. ........ 19
*Kunal Velinkar, Alex Von Gunten, Jeffrey Greeley, Eranda Nikolla*

212f Understanding the Chemistry and Formation of Fsi-Derived Solid-Electrolyte Interphase at Li-Metal Anode ........ 20
*Weilai Yu, Yi Cui, Zhenan Bao*

212g Interface Engineering in Low-PGM Catalysts for Direct Ethanol Fuel Cells ........ 21
*Yang Yang*

212h Ion Transport and Interfacial Dynamics in Polymerized Ionic Liquids ........ 22
*Javad Jeddi, Joshua Sangoro*

212i Tuning Au@Pd Nanoparticles Interfacial Engineering for Plasmon-Enhanced Hydrogen Evolution Performance ........ 23
*Lemma Tufa*

## MATERIAL INTERFACES FOR CLEAN ENERGY: FUNDAMENTALS AND APPLICATIONS

126b Modeling of Spatiotemporal Temperature Distribution in Hybrid Nanoscale Multifunctional Material for Desorption of $CO_2$ ........ 24
*Raj Panthesh Bhavsar, Bhavana Bhadriraju, Ga Hyun Lee, Ah-Hyung Park, Joseph Kwon*

126c Structure and Transport Behaviors of Liquid-like Nanoparticle Organic Hybrid Materials Designed for Combined $CO_2$ Capture and Conversion ........ 26
*Ga Hyun Lee, Tony Feric, Sara Hamilton, Xiaoyang Shi, Ah-Hyung Park*

126e Ultrathin: Understanding the Nanostructure of Alumina Atomic Layer Deposited Films on Layered Cathode Oxides ........ 27
*Julie A. Nguyen, Elizabeth K. Allan-Cole, Meghan N. Herbert-Walters, Madison B. Martinez, Barbara K. Hughes, Michael F. Toney, Hendrik Heinz, Alan Weimer*

126f Molecular Design of Electrochemical Interfaces for Sustainable Separations in Clean Energy ........ 28
*Xiao Su*

126g Novel Electrocatalysts for Nitrogen Oxides Reduction at Ambient Conditions ........................................ 29
*Feng Jiao*

## MATERIALS INTERFACES - CELEBRATING THE CAREER OF JOHN EKERDT

24a Chemical Processing of Function Accelerated Nano-Materials at the Atomic Scale .................................. 30
*Jane Chang*

24b Surface Reaction Kinetics during Area-Selective Thin Film Deposition .................................................. 31
*Gregory N. Parsons*

24c Controlling Nucleation and Selectivity in Atomic Layer Deposition for Materials Synthesis..................... 32
*Stacey Bent*

24d High Quality Tunable Plasmonic Nanostructures By Atomic Layer Deposition......................................... 33
*Brian Willis, John A. Grasso, Rahul Raman*

24e Chemical "Engineerifying" Electronics............................................................................................. 34
*Michael A. Filler*

24f The Elusive Surface M2O7 Dimer Active Site .......................................................................................... 35
*Israel Wachs*

24g The Dual Nature of Chemical Reaction Engineering ................................................................................. 36
*James Rawlings*

24h Plasma Enhanced Atomic Layer Deposition of Silicon Carbonitride....................................................... 37
*Samuel Johnson, Tsung-Hsuan Yang, Jianping Zhao, Toshihiko Iwao, Charles Schlechte, John Carroll, Gabriel Blankemeyer, Peter Ventzek, Joaquin Resasco, Gyeong S. Hwang, John Ekerdt*

**Author Index**

# 335a Elucidating Non-Thermal Plasma-Activated Carbon-Nitrogen Coupling on a Catalytic Surface from Methane and Nitrogen

*Garam Lee, David Go, Casey O'Brien*

From inelastic collisions between the energetic electrons and gas molecules, non-thermal plasma (NTP) produces reactive chemical environments (e.g., electrons, ions, radicals, and vibrationally-excited molecules) at temperatures where such species are thermally inaccessible. Integrating reactive NTP with heterogeneous catalysis has recently gained interest to enhance catalytic activities/selectivities and promote a novel chemical transformation that neither plasma nor catalysis could deliver individually. To better control plasma-catalyst coupling, it is important to understand the interaction between the NTP-activated species and a catalytic surface. In this work, we utilized a multi-modal spectroscopy tool combining polarization-modulation infrared reflection-absorption spectroscopy (PM-IRAS), optical emission spectroscopy (OES), and mass spectrometry (MS) to investigate the interaction between NTP-activated $CH_4/N_2$ species and model surfaces (Ni, Pd, Cu, Ag, Au, and $SiO_2$) for C-N coupling. To minimize the plasma-phase interaction between $CH_4$ and $N_2$, we introduce $CH_4/N_2$ feed gas sequentially: (1) $CH_4$-$N_2$ in sequence and (2) $N_2$-$CH_4$ in sequence. We show that different sequences lead to CN signatures in OES spectra corresponding to C-N coupled products forming on a catalytic surface. The effects of different surfaces on the nature of surface-adsorbed C-N coupled products are further investigated with ex-situ X-ray photoelectron spectroscopy (XPS) and liquid chromatography-mass spectrometry (LC-MS).

## 335b Pathways and Timescales for $N_2$ Conversion By Reductive or Oxidative Routes in Plasma Catalysis

*Brian Bayer, Peter Bruggeman, Aditya Bhan*

We report pathways and timescales for reductive and oxidative conversion of plasma-derived $N_2$ species (N, $N_2(v)$) using a combination of molecular beam mass spectrometry (MBMS) measurements to enumerate densities of plasma-derived species and short contact time reactors with sub-millisecond residence times. For reduction of $N_2$ in $Ar/N_2/H_2$ mixtures, rates and quantities of $NH_3$ formation correlate with rates and quantities of N consumption, indicating that $NH_3$ formation occurs from surface-mediated reactions involving N radicals. When densities of H and $H_2$ are sufficient, conversion of N to $NH_3$ is 100% selective over Fe, Ni, and Ag. Through MBMS-validated state-to-state vibrational kinetic modeling, we show that $N_2(v)$, though produced in quantities exceeding N by 100×, is not reactive for $NH_3$ formation at the investigated experimental operating conditions and loss of $N_2(v)$ occurs due to vibrational relaxation on the catalyst surface. For oxidation of $N_2$ in $Ar/N_2/O_2$ mixtures, the timescale for N consumption in the gas phase decreases from ~$10^{-3}$ to ~$10^{-4}$ s as the $O_2$ concentration increases from 0.1-5%, with gas-phase NO formation increasing as the $O_2$ concentration increases. When catalyst is present in the reactor, the timescale for N consumption decreases, and MBMS measurements show that NO formation with Fe or Ag catalyst exceeds NO formation from gas-phase reactions alone at equal residence time, although NO formation is always less than N consumption. These findings and timescales for N consumption in the gas phase and with catalyst present in oxidative and reductive environments will be discussed.

## 335c Elucidating Plasma-Catalyst Contributions for the Direct Conversion of Methane to Aromatics

*Gerardo Rivera-Castro, Ibukunoluwa Akintola, David Go, Jason Hicks*

The direct transformation of methane to liquid fuels and chemicals remains a critical research target due to the availability of natural gas and the abundance of methane in these resources. Methane dehydroaromatization is a promising route for the direct conversion of methane to aromatics (i.e., benzene, toluene, xylenes) under non-oxidative conditions. Among active metals studied for this reaction, Mo/H-ZSM-5 has been shown to be efficient toward C-H activation and formation of aromatic compounds. However, the commercialization of this process remains unfeasible, mainly due to the endothermic nature of this reaction ($\Delta H_R{}^0 = + 532$ kJ/mol), where equilibrium conversion is limited to 14% at 973 K. In addition, these conditions favor coke formation over the surface of the catalyst, leading to rapid catalyst deactivation. An alternative to the thermal conversion of methane to aromatics is the use of non-thermal plasma technologies. Non-thermal plasmas can facilitate chemical conversions of stable molecules (i.e., $CH_4$, $CO_2$ and $N_2$) at ambient pressure and temperatures through electron-impact reactions. Therefore, combination of the dehydroaromatization catalyst and non-thermal plasmas could potentially mitigate the limitations presented for this process.

In this work, we have addressed questions that remain unanswered regarding the role of the active metal (i.e., Mo) for the conversion of methane to aromatics in a one-pot dielectric barrier discharge reactor. We have specifically investigated the influence of the bulk gas temperature of the plasma-assisted reaction to elucidate the functionality of Mo and determine the contribution to plasma-phase reactions and Mo-catalyzed reactions. Our work shows that conversion of methane to aromatics can take place at temperatures as low as 573 K, where the multifunctionality of Mo/H-ZSM-5 is not required under these conditions. As the bulk gas temperature for the reaction is further increased, we observe additional contributions from Mo-catalyzed pathways toward the production of aromatics under plasma stimulation. This result is supported by the observation of Mo carbide species that are obtained at temperatures of 773 K in the presence of the plasma, which are not observed under thermal conditions using methane as a source. Overall, this presentation will highlight our current understanding of the role of plasma and catalyst for the conversion of methane to aromatics, and the influence of the bulk gas temperature over these complex plasma-catalyst interactions.

# 335d Machine Learning Boosted Design of Plasma-Enhanced Catalytic Ammonia Cracking

*Saleh Ahmat Ibrahim, Hsi-Wu Wong, Juan Pablo Trelles, Fanglin Che*

Hydrogen ($H_2$) is an energy carrier with high energy content (142 kJ/g) and a non-polluting energy source. Hydrogen's implementation is hindered, however, by delivery and storage issues. Ammonia ($NH_3$) is being studied as a storage medium for $H_2$ due to its high $H_2$ density, ability to easily condense, storage and distribution. However, thermal $NH_3$ cracking to generate $H_2$ usually requires the reaction temperatures above 773 K and an earth-rare metal catalyst (i.e., Ru) to achieve complete conversion.

One strategy to activate ammonia at mild conditions over earth-abundant metal catalysts is by applying non-thermal plasma since the vibrationally excited species produced in non-thermal plasmas lower the temperature of $NH_3$ cracking. However, modelling interfacial plasma-surface interaction is a multi-scale problem and building such simulation for a complex reaction mechanism over a large search space of catalyst materials is computational immense.

To solve the above challenge, it is important to identify the key descriptor that impacts the overall rates of ammonia cracking under non-thermal plasma conditions and apply such descriptor to screen the large search space of catalysts. Thus, we first built the microkinetic model for ammonia cracking over various metallic surfaces (e.g., Ni, Co, Ru, Fe). Our theoretical results indicate that plasma lowers the energetics of the first N-H bond cleavage in NH3 and improves the overall rates by creating vibrationally excited state. Then we performed the sensitivity analysis to identify the key descriptor of $N_2$ dissociative desorption from surface $N^*$. We then applied this descriptor to scan the catalyst materials and found that earth-abundant Ni-Co alloy showed satisfactory catalytic performance compared to that of earth-rare Ru. AI-predicted catalysts will be further validated by experiments using a dielectric-barrier discharge reactor. The integration of multi-scale modeling, machine learning algorithms and experiments will facilitate materials development in interfacial plasma-surface reaction rather than trial-and-error.

## 335e Selective Propylene Epoxidation from Plasma-Liquid Interactions Using Water As the Oxygen Source

*Dongho Lee, Han-Ting Chen, Suljo Linic*

Propylene oxide (PO) is a critical feedstock chemical for the large-scale production of polyurethane plastics and other compounds. While centralized and large-scale PO production has its advantages, due to the high toxicity of PO and the high transportation and long-term storage costs, there is also a strong incentive to produce PO on-site and in time on a smaller scale for applications that do not necessarily require a high volume and concentration. These niche applications include using PO as a fumigant and disinfectant for food,[1] a sterilizer for medical equipment,[2] and in forming modified food such as starch and alginate.[3] Herein, we designed a plasma-liquid interaction catalytic process using only water and propylene ($C_3H_6$) as reactants to form PO. Plasma-liquid interactions can activate a wide range of chemical transformations at ambient conditions and form hydrogen peroxide ($H_2O_2$) with relatively high efficiencies.[4, 5] The $H_2O_2$ generated in the plasma-liquid interactions serves as an oxidizing agent that epoxidizes the dissolved $C_3H_6$ over a titanium silicate-1 (TS-1) catalyst dispersed in the water solution. We explored different reactor designs to suppress the generation of deep oxidation products (i.e., $CO_2$) and reach a carbon-based selectivity of more than 98%. Also, the influence of plasma parameters on the activity and selectivity of PO production is investigated. The findings provide us with a method to produce on-demand PO using only $C_3H_6$ and water, which can be suitable for various specialized applications.

References

1. Danyluk, M. D.; Uesugi, A. R.; Harris, L. J., Survival of Salmonella enteritidis PT 30 on inoculated almonds after commercial fumigation with propylene oxide. *J Food Protect* **2005,** *68*, 1613-1622.
2. Sato, H.; Kidaka, T.; Hori, M., Sterilization of Therapeutic Immunoadsorbents with Aqueous Propylene-Oxide Solution. *Int J Artif Organs* **1985,** *8*, 109-114.
3. *Propylene Oxide - Environmental Health Criteria 56*; Finland, 1985.
4. Bruggeman, P. J.; Frontiera, R. R.; Kortshagen, U. R.; Kushner, M. J.; Linic, S.; Schatz, G. C.; Andaraarachchi, H.; Exarhos, S.; Jones, L. O.; Mueller, C. M.; Rich, C. C.; Xu, C.; Yue, Y. F.; Zhang, Y., Plasma-Driven Solution Electrolysis. *J Appl Phys* **2021,** *129*, 200902.
5. Wang, H. H.; Wandell, R. J.; Tachibana, K.; Vorac, J.; Locke, B. R., The Influence of Liquid Conductivity on Electrical Breakdown and Hydrogen Peroxide Production in a Nanosecond Pulsed Plasma Discharge Generated in a Water-Film Plasma Reactor. *J Phys D Appl Phys* **2019,** *52*, 075201.

## 335f Decoupling Plasma-Catalytic Interactions through Plasma-Temperature Programmed $N_2$ Reduction and Oxidation Reactions

*Hope Otor, Patrick Barboun, Hanyu Ma, Anshuman Goswami, William Schneider, Jason Hicks*

The potential for non-thermal plasma to activate strong chemical bonds has provided alternative routes for chemical transformations at milder reaction conditions. Plasma-activated catalysis combines plasma and catalytic contributions to enhance chemical reactions. The choice of catalytic material is pivotal in harnessing the synergistic relationship observed when stimulated by plasma. However, material design that can effectively harness the plasma-catalyst synergy depends on the fundamental understanding of relevant plasma-activated surface reaction pathways. Moreover, in these plasma-catalytic systems, plasma phase reactions can dominate, which often complicates the analysis and contribution of the catalyst surface reactivity to the overall product yields.

This presentation will focus on our recent investigation in elucidating the reactivity of plasma-activated nitrogen species on a metal surface. To do this, we employ sequential exposure of the catalyst surface to the different reactant gases in which the plasma-activated nitrogen is subjected to temperature-programmed reactions with another reactant (hydrogen or oxygen). By so doing, we can evaluate the reactivity of plasma-activated nitrogen under reductive and oxidative conditions. Under hydrogenation conditions, we observe the formation of $NH_3$ over different silica-supported metal catalysts (Fe, Co, Ni, and Pt). Hence, this shows that after plasma activates nitrogen, metal surfaces can perform hydrogenation chemistry to produce $NH_3$. Further, microkinetic modeling calculations show that the temperature of $NH_3$ formation scales with the activation barrier for surface hydrogenation. Under oxidative conditions over Pt and Au catalysts, we observed the formation of NO and $N_2O$. Interestingly, we observed that the metal affinity towards oxygen activation is critical in directing product selectivity, with Pt being more selective to NO while Au is more selective to $N_2O$. This further elaborates on the effect of metal surface reactivity under plasma stimulation. Taken together, our study highlights the significance of catalytic contribution and surface reactivity, which is highly relevant for guiding material selection and design principles for plasma-activated processes.

# Mn-Ferrite As a Highly Active Catalyst for Toluene Degradation in Post-Plasma Catalysis

*Khurram Shahzad Ayub, Waqas Qamar Zaman, Zain Abbas*

Volatile organic compounds (VOCs) can pose a significant health risk to humans even at low concentrations. Non-thermal plasma (NTP) is an alternative technology to conventional thermo-catalytic methods for treating low-concentration VOCs. However, its high energy consumption and the production of toxic organic byproducts limit its application. To address these limitations, combining NTP discharge with catalysis has emerged as a promising technique which is known as post-plasma catalysis (PPC). This study aims to design energy-efficient and high-performance catalysts that are best suited to the plasma conditions. Toluene is chosen as the target VOC for decomposition in a dielectric barrier discharge plasma reactor at various input power levels, and a series of Mn-ferrites catalysts with different Mn/ferrite molar ratios are used to investigate their synergistic effects on the decomposition efficiency of toluene. The results demonstrate that the Mn/Ferrite molar ratio influences the physicochemical properties of the catalyst due to its catalytic activity. Nonetheless, PPC with all catalysts enhances the decomposition efficiency of toluene at all input power levels when compared to NTP-alone. Among all catalysts, Mn1-Ferrite1 exhibits the highest decomposition efficiency of toluene at all input power levels while reducing unwanted organic products' formation. These findings promote the practical application of PPC integrated with bimetallic oxide catalysts.

## 335g Evaluate the Feasibility of Employing Cactus Mucilage to Improve the Robustness of the Reactive Coating Photocatalyst for the Removal of Chosen Bacteria in Wastewater

*Khuloud Alrashdi, Norma Alcantar*

The presence of wastewater creates a perfect habitat for the development of a broad range of bacteria and viruses, including Escherichia coli (E. coli), Salmonella, and Rotavirus. The bacterium Escherichia coli, sometimes known as E. coli, is commonly found in both human and animal digestive systems. When E. coli is identified in wastewater, it may indicate that the water has been polluted with excrement, which may lead to health problems. Due to its extensive dispersion, the bacterium Staphylococcus aureus (S. aureus) may be found in wastewater. S. aureus is the causal agent of a broad range of infections, from those that just affect the skin to those that may be deadly, such as pneumonia and septic shock. Salmonella is a pathogen that may be transferred from animals to people via the eating of infected food. Salmonella enterica is the most frequent kind of bacterium that causes illness, although there are many others. It has also been identified in polluted food and water. Salmonella may be found in high numbers in the wastewater of certain kinds of institutions, such as hospitals, nursing homes, and other facilities that house a significant number of animals, such as farms and slaughterhouses.

Photocatalysis is rapidly becoming one of the most popular technologies for wastewater treatment due to its low cost, environmental friendliness, and adaptability. There are numerous different technologies for cleaning wastewater, but photocatalysis has significant features that set it apart. This wastewater treatment technique is not only cost-effective, but it can also be readily scaled up to handle large volumes. There is no need to use costly or perhaps harmful chemicals. The application of this approach has the potential to clear the environment of microplastics, medications, and other developing types of pollution.

Mucilage is a viscous, gel-like material present in many plants, including aloe vera, cactus, and okra. Because of its unique features, such as its ability to lower surface tension and stabilize emulsions, mucilage has been investigated as a possible alternative to synthetic surfactants. Surfactants are substances that may lessen the surface tension between two liquids or between a liquid and a solid, making it easier for them to mix and interact. They are frequently utilized in a broad range of applications, such as cleaning, foaming, and emulsification. The use of mucilage as a surfactant for photocatalysts has the potential to increase photocatalytic reaction efficiency. Photocatalysts are materials that may enhance chemical reactions when exposed to light, and they are commonly employed in wastewater treatment and air purification to degrade contaminants.

In this work, we studied the possibility of spray coating photocatalyst onto a substrate with cactus mucilage to improve the chemical structure of the photocatalyst. Titanium dioxide ($TiO_2$) doped Zinc oxide thin films were produced onto glass substrates utilizing a spray coating method at various annealing temperatures and concentrations. X-ray diffraction (XRD), Fourier transform infrared (FTIR), UV-Vis diffuse reflectance spectroscopy, scanning electron microscopy (SEM), and transmission electron microscopy were used to analyze the new photocatalyst (TEM). S degradation was used to assess photocatalytic activity. aureus, as well as E. coli

## 257a Physics and Data-Informed Formulation Design and Development for Chemical Processes

*Shikha Mallick, Francesca Boioli, Ravindra Aglave, Panos Petris, Peter Mas*

The design of novel material formulations is a critical innovation process. The shelf-life and the performance of the final product depends on fast and informed decisions adopted early in the design phase. Digital ways to achieve this can not only accelerate the project timescales but reduce repetitive tasks and errors. Formulation management has been digital for a a while now in the industry. However, it has been done disjointed with the deep chemistry related insight necessary to innovation.

The use of molecular scale modeling is a method that has been used as well in the industry but not integrated into the formulation management workflows. In this work we present how the connection between formulation management and chemistry innovation can be carried out in the context of innovation and innovation project management.

For this purpose, we take solubility as a variable, critical in many aspects of formulation development. As a first step, a limited set of molecules from a specific chemical family (or families) is selected, and their solubility is calculated using molecular simulations. More specifically we estimate the Hansen solubility parameters[1] using a fast Molecular Dynamics technique, where the simulation box is successively compressed and de-compressed to reach faster in correct density and thermodynamic equilibrium.[2] The solubility parameters are defined by the intermolecular energy of the simulation box.

However, to reduce further the timescales of development, we combined Graph Neural Network (GNN) architectures with physics-based simulations to estimate materials properties of industrial formulations. The adopted machine learning model is very meticulously selected. Traditional machine learning algorithms and deep learning algorithms[3] were also tested and provided high accuracy results, but molecular descriptors had to be generated to train the model against them (i.e. asphericity, molecular weight, number of hydrogen bonding donors/acceptors, radius of gyration, etc). The advantage of GNNs is that the complex chemical structure of a molecule can be naturally represented by a graph, defined as the ensemble of the connectivity relationships between a set of nodes (atoms) and a set of edges (bonds).[4] That allows us to immediately train against the target solubility parameters without additional calculations.

This information can now be made available in the formulation management digital tools. Physics and data-informed materials property prediction in milliseconds is essential for an agile formulation development and decision-making. It allows for the exploitation of the vast chemical space for the optimal chemistry, something that cannot even be remotely accomplished by physical testing in relevant industrial cost and time scales.

(1) Hansen, C. M., J. Paint Technol. 1967, 39, 511

(2) Belmares, M.; Blanco, M.; Goddard, W. A.; Ross, R.; Caldwell, G.; Chou, S.-H.; Pham, J.; Olofson, P. M.; Thomas, C., J. Comput. Chem. 2004, 25: 1814–1826

(3) Deng, T.; Jia, G.-Z., Molecular Physics, 2020, 118:2

(4) Lee, S.; Lee, M.; Gyak, K.-W.; Kim, S.; Kim, M.-J.; Min, K., ACS Omega 2022, 7, 14, 12268–12277

## 257b Data-Driven Optimization of Perovskite Solar Cells

*Donghyun Oh, Sanggyun Kim, Carlo Andrea Riccardo Perini, Juan-Pablo Correa-Baena, Nikolaos Sahinidis*

Perovskite Solar Cells (PSCs) have gained significant attention in the field of photovoltaics in recent years. While lab-scale PSCs have achieved impressive power conversion efficiencies exceeding 25%, there is a lack of systematic and rigorous strategies to optimize various parameters affecting their performance. For instance, material composition and fabrication process variables interact in complex ways, thus presenting a significant challenge for search algorithms. At present, researchers primarily depend on extensive trial-and-error to explore an enormous combinatorial space, which can be both time-consuming and costly, and may lead to suboptimal results.

In this work, we present a combined experimental-computational framework for efficient development of high-performing PSCs. In terms of experimentation, we fabricate organic–inorganic halide perovskite based solar cells via solution processing. We extract photovoltaic parameters from J-V characteristics measured under illumination by a solar simulator. On the computational side, we use black-box optimization methods to guide the search for optimal designs. These algorithms construct models to capture the impact of different factors affecting device performance and predict the desirable combination of design variables. Our results demonstrate that the proposed approach enables rapid and systematic development of high-performing solar devices.

## 257c Controlling Structure-Property Relationships in Organic Semiconductors Using Tunable Highly-Ordered 2D Perovskite Interfaces

*Meagan Phister, Ashley Conley, Rachel Dziatko, Karl Westendorff, Anna Sviripa, Stephanie Guthrie, Kevin Stone, Arthur Bragg, Christopher Paolucci, Joshua Choi, Gaurav Giri*

Organic Semiconductors (OSCs) have been extensively studied in a variety of fields such as device engineering and solid-state physics for their optical and electronic properties. OSCs are solution processable, which allows for high-throughput fabrication of lightweight, flexible devices, making them a competitive alternative to amorphous and polycrystalline silicon semiconductors. A challenge of utilizing OSCs is their natural tendency to crystallize into multiple molecular packing motifs, called polymorphs, which yield highly variable charge transport capabilities and optoelectronic properties. Also affected is the property of singlet fission, an exciton multiplication process in which one absorbed photon is converted into two triplet excitons. This sharing of energy from one photon to two triplet excitons could result in quantum efficiencies greater than 100% providing an opportunity to surpass the Shockley-Quisser limit and lead to the creation of higher efficiency solar cells.

In this work, two-dimensional (2D) lead halide perovskites were utilized as a highly ordered self-assembled monolayer (SAM) with tunable ligand density and variable ligands, in which six interface-stabilized polymorphs of a small molecule OSC, namely TIPS-pentacene were isolated via meniscus guided coating and subsequent thin film relaxation in-order to release strain. These interface-stabilized polymorphs were identified using grazing incidence x-ray diffraction, where incremental changes in the (101) plane spacing were elucidated, having a small range of 0.12Å. Distinct bandgap transition energies and relative deviations in steady-state photoluminescence were discerned, indicating varying degrees of intermolecular interactions arising from the unique crystal structures. Through our work, we study the effect of crystal structure on singlet fission without introducing the effects of crystal morphology, altering the polymorph, or relying on metastable structures. Singlet fission rates were enhanced due to the increased order and control of TIPS-pentacene molecular packing as a result of creating an interface-stabilized structure. This demonstrates that for singlet fission, long-range molecular coherence has a significant impact on the resulting rate. This work motivates the use of a variety of combinations of 2D lead halide perovskites and small molecule organic semiconductors to optimize and control structure-property relationships utilizing interface engineering.

## 257d Benchmark Simulation of Ag Nanoparticles on Supported Graphene Substrates

*Tristan Maxson, Tibor Szilvasi*

Nanoparticles have a wide range of applications (thermal-, electro-, and photocatalysis, sensors, etc.) but their in-situ characterization and computational modeling prove challenging. Current nanoparticle models typically use low-index periodic models with defect-free surface due to computational cost reasons. The recent development of machine learning force fields (MLFFs) provide computationally viable approximations of density functional theory calculations while maintaining similar levels of accuracy in energies and forces to model nanoparticles as is. To benchmark the validity of MLFFs for nanoparticle simulations, we build upon the recent work of Campbell et al[1] that gives an experimental understanding of the adhesion of Ag nanoparticles to Ni(111) supported graphene and their chemical potential as a function of particle size. In this work, we train MLFFs using Allegro via an iterative committee training scheme to identify poorly predicted structures and improve the training set. Hybrid Grand Canonical Monte Carlo and Molecular Dynamics simulations are performed to explore favorable geometries of nanoparticles at various temperatures to make comparison to Campbell's results and also compare the resulting optimal nanoparticle shapes with nanoparticles formed via the Winterbottom construction. This work marks the first progress toward understanding how supported nanoparticles look like under experimentally relevant conditions as we move from the nanocluster (diameter below 1 nm) to the nanoparticle size regime (diameter above 1 nm).

(1) Rumptz, J. R.; Mao, Z.; Campbell, C. T. Size-Dependent Adsorption and Adhesion Energetics of Ag Nanoparticles on Graphene Films on Ni(111) by Calorimetry. *ACS Catalysis* **2022**, *12* (5), 2888-2897.

## 257e Effects of the Polymer/MOF Interface in Mixed-Matrix Membranes on Small Molecule Transport

*Moein Hajian Z.D., Hyunhee Lee, Zachary P. Smith, Joshua Howe*

Despite high interest in membrane-based separations, sometimes pure polymeric membranes do not possess the necessary selectivity and permeability required for certain applications. Mixed-matrix membranes (MMMs) incorporate filler particles into polymer matrices and present a route to improving selectivity without sacrificing permeability to enhance separation performance. Metal-organic frameworks (MOFs) are common filler particles used in MMMs due to their chemical diversity and tunable molecular interactions at well-defined adsorption sites. While there has been substantial experimental work in this area in recent decades, theoretical developments in this area have lagged behind due to key challenges in accurately characterizing MMM materials. While the length- and time-scales relevant to polymer physics typically involve hundreds of thousands of atoms and slow relaxations, accurate characterization of MOFs often relies on electronic detail using methods incapable of robustly describing polymer materials. To address this challenge, we have developed models of polymers that do not fully capture polymer behavior but do allow for the study of the electronic effects at the interface to see how the presence of polymers may affect interactions and transport at the polymer/MOF interface in MMMs [1].

In this work, we study the diffusion of hydrogen, nitrogen, methane, propane, propylene, and iso-butane through two facets (100 and 110) of zeolitic imidazolate framework-8 (ZIF-8) with unsaturated metal sites on the surface, both pristine and in contact with models of 6FDA-DAM. We use density functional theory with climbing image nudge elastic band calculations [2] to investigate the electronic and structural properties of the MOF-polymer interface during the transport of small molecules across the polymer/MOF interface in MMMs. The study aims to gain insights into the electronic interactions of MOF-polymer interfaces and their impact on gas separation performance. We compare our computational calculations with an experimental study that characterizes the diffusion of gases through the selected MMM. This work establishes insights into the structural and electronic properties of MOF/polymer interfaces and evaluates the energy barriers emergent at these interfaces that may affect transport and separation properties in MMMs. We envision that these insights can be used to design new materials with improved performance by establishing design principles for optimizing the polymer/MOF pairings in MMMs.

[1] Sadeghi, Sina, and Joshua D. Howe. 'Ab Initio Study of Adsorption of Polymers on Metal–Organic Framework Surfaces.' *The Journal of Physical Chemistry C* (2023).

[2] Henkelman, Graeme, Blas P. Uberuaga, and Hannes Jónsson. 'A climbing image nudged elastic band method for finding saddle points and minimum energy paths.' *The Journal of chemical physics* 113, no. 22 (2000): 9901-9904.

## 257f Utilizing Tunable Surfaces for Performance and Stability Enhancement of Nitramine Energetic Materials

*Natalie Smith, Meagan Phister, Gaurav Giri*

The development of novel and high performing energetic materials is a continuous focus for defense applications, as well as for propellants, construction, and pyrotechnics, amounting to a multibillion-dollar industry. Nitramines are a class of secondary explosives comprised of organic molecules containing nitro groups bonded to a nitrogen atom (e.g., TNT, HMX, RDX, CL-20). Compared to other nitramines, CL-20 has higher energy density, detonation velocity, and an enhanced oxidizer-to-fuel ratio; however, limitations relating to its high impact sensitivity and facile phase transformations hinder its implementation commercially.

To enhance stability and performance, energetic crystals are often dispersed in polymer matrices, most commonly hydroxyl terminated polybutadiene (HTPB). Interactions between the polymer binder and the crystal surface are difficult to study and incompatible interactions between the two components could result in delamination and crystal phase transformation, ultimately leading to unpredictable stability and performance. A better understanding of how energetic crystals behave when interfaced with various chemical functionalities is crucial for safer and higher performing energetic composites.

In this work, thin films of CL-20 are crystallized via meniscus guided coating to enable facile observations at the interfacial boundary between EM and composite materials. CL-20 thin films are interfaced with a highly tunable 2D metal-halide perovskite (MHP) surface. 2D MHPs assemble to generate uniform surfaces with tunability of exposed chemical group functionality and density. X-ray diffraction, Raman spectroscopy, and optical microscopy are employed to characterize the EM crystal structure, strain, and morphology, respectively. Interfacial interactions influence the density of EM packing, the strain induced within the crystal structure, and ultimately the stability of the EM crystals.

## 257g Photochemical Bubble Generation on Polymer-Water Interface.

*Pranaya Ghate, Christopher J. Bardeen, Rabih O. Al-Kaysi*

The ability to control solid-gas phase transitions using light as a stimulus has attracted significant interest in the field of materials science because it gives rise to a large volume change. In most cases, this transition is caused by photothermal heating of a volatile liquid. In this work, we use a photochemical mechanism for gas generation that is based on using organic azides undergoing UV photolysis to release gaseous nitrogen.

2-Anthraceneazide is doped into a 150μm thick polymer films that release nitrogen microbubbles on the film surface when irradiated in a water reservoir. The formation of nitrogen bubbles was observed using optical microscopy, and their size and distribution were characterized using image analysis. The effects of various irradiation conditions, such as irradiation time and light intensity, doping concentration of azide, and film surface on the formation of nitrogen bubbles were investigated. The results indicate that the size and distribution of the bubbles can be controlled by adjusting the irradiation conditions and film surface morphology.

The polymer film containing 2-anthracene azide, can be floated to the surface by the photoinduced bubble coating. The inherent property of a bubble is the air-water interface that attenuates ultrasound waves, making it a desirable option to use for acoustic blocking. Photochemical bubble generation provides a new avenue for the development of materials and devices that can be controlled by light and has potential applications in areas such as microfluidics, acoustic blocking, and nanotechnology.

## 212a Investigating the Coupled Effect of IFT Reduction and Wettability Alteration in Enhanced Oil Recovery from Carbonate Reservoirs

*Muhammad Shahzad Kamal, Deng Xiao, Shirish Patil, Syed Hussain*

Carbonate reservoirs are often oil-wet and naturally fractured, leaving about 2/3 of the oil Carbonate reservoirs often retain around 2/3 of their oil after secondary recovery, necessitating the use of Enhanced Oil Recovery (EOR) techniques to improve oil recovery efficiency. Two of the most crucial mechanisms for EOR in carbonate reservoirs are Interfacial Tension (IFT) reduction and Wettability Alteration. Although both methods contribute to oil recovery when used separately, their combined effect can lead to confusion about whether they have a synergistic or offsetting impact. Some studies suggest that IFT reduction can enhance wettability alteration. In this research, the study compares the performance of wettability alteration with and without IFT reduction and concludes that IFT reduction has a negligible effect on wettability alteration. The research employed Indiana limestone outcrops, filtered crude oil, a commonly used chelating agent (DTPA), and a zwitterionic surfactant (ZS). The limestone outcrops were cut into 1-inch diameter and 3-4 mm thickness slices, smoothened using P500 sandpaper, and then aged in crude oil at 90°C. Different concentrations of DTPA solutions were diluted using synthetic seawater, and ZS was added to some solutions, while some only had ZS seawater solutions without DTPA. The IFT values between oil and solutions were measured, and the oil-aged slices were treated with different solutions to reflect on their wettability alteration performance. The IFT results showed that ZS is highly effective in reducing IFT values down to 0.01 mN/m, while having a negligible effect on wettability. On the other hand, DTPA is highly effective in changing rock from strongly oil-wet to water-wet conditions but has limited impact on IFT. When DTPA and ZS were combined, the solutions were effective in both reducing IFT and changing wettability. However, the wettability alteration performance of DTPA+ZS was similar to that of DTPA alone, indicating that even significant IFT reduction has negligible influence on the wettability alteration performance.

## 212c Chemical Oscillation at the Grain Boundary of Aliovalently-Doped Solid-Oxide Electrolytes

*Bharathi Bandi, Abhijit Chatterjee*

The ionic conductivity of polycrystalline aliovalently-doped solid oxide electrolytes is generally lower than that of the corresponding single-crystal material by a few orders of magnitude. This behavior arises due to slow ionic movement in the space-charge region at the grain boundary (GB).Although traditionally the Guoy-Chapman (G-C) and Mott-Shottky (M-S) models have been used to explain the nature of the space charge layer and ionic conduction at GB, these models ignore the chemical interactions and the molecular scale-atomic structure at the GB. To highlight the importance of chemical interactions and GB misorientation, we systematically probe the space-charge layers at Σ5 and Σ13 GBs in two commonly-used solid electrolyte materials, namely, yttria stabilized zirconia (YSZ) and gadolinium doped ceria (GDC). Using molecular dynamics (MD), we provide first evidence of chemical oscillation in the O2- ion concentration profile in the space charge region. These oscillations arise from the competing chemical, electrostatic and elastic interactions, which is missed by the G-C and M-S models. Space charge layer characteristics in each of the GBs studied is found to be different. An alternate theory which explains the chemical oscillations in the space-charge layer is presented. Implications on the ionic conduction are discussed

## 212e Impact of Cathode/Solvent Interfaces on the Electrochemical Performance of Na-$O_2$ Batteries.

*Kunal Velinkar, Alex Von Gunten, Jeffrey Greeley, Eranda Nikolla*

Aprotic Na-$O_2$ batteries have recently gained significant attention as promising alternatives to commercial Li-ion batteries owing to their high energy densities and reversible redox chemistries[1–3]. The underlying electrochemistry in these systems is heavily influenced by the glyme-ether electrolyte solvents owing to their solvation properties that vary as a function of their respective chain lengths[4]. Na-$O_2$ systems with short chain ether electrolyte solvents are challenged by their long-term stability[5] whereas, the systems with long chain ether electrolyte solvents lack fundamental understanding in their underlying electrochemistry[1]. These challenges hamper the development of effective strategies to improve their stability and overall electrochemical efficiency[3]. Herein, we investigate the effect of the cathode/aprotic electrolyte solvent interface on the mechanism of discharge product formation and overall electrochemical cell performance by varying both the glyme-ether electrolyte solvent and the carbon electrode. Theoretical calculations are coupled with experimental X-ray diffraction (XRD), scanning electron microscopy (SEM), and X-ray photoelectron spectroscopy (XPS) studies to determine the nature of the Na-$O_2$ species formed as a function of the electrolyte solvent and carbon cathode. A link between the nature of the carbon cathode/aprotic solvent interface, discharge mechanism for product formation and the overall electrochemical performance and stability is reported. These insights are critical for the rational design of efficient and stable Na-$O_2$ batteries.

**References:**

(1) Samira, S., et al., *ACS Energy Lett,* **2021**, *6* (2), 665–674.

(2) Hartmann, P., et al., *Nat. Mater.* **2013**, *12* (3), 228–232.

(3) Bi, X., et.al., *Small Methods,* **2019**, *3* (4), 1800247.

(4) Lutz, L., et.al., *The Journal of Physical Chemistry C,* **2016**, *120* (36), 20068–20076.

(5) Kim, J., et.al., *Nat Commun,* 2016, *7* (1), 10670.

## 212f Understanding the Chemistry and Formation of Fsi-Derived Solid-Electrolyte Interphase at Li-Metal Anode

*Weilai Yu, Yi Cui, Zhenan Bao*

Li metal battery (LMB) has always been considered as the holy grail to break the current limitations in energy density of Li-ion battery. However, commercialization of LMB has long been plagued by numerous operational issues such as Li dendrite formation causing safety concerns, as well as uncontrolled interfacial reactivity leading continuous electrolyte consumption.

Developing new liquid electrolyte based on ether solvent and LiFSI salt is a promising strategy of enhancing the reversibility of Li-metal anode (LMA), with the state-of-the-art values of Coulombic efficiency (CE) exceeding 99.5 %. Nevertheless, the chemical origin of Coulombic inefficiency (CI) and its correlation with the formation of solid-electrolyte interphase (SEI) at Li metal surface have not been well understood. A fundamental look at these basic processes accompanying the continuous cycling of LMA, can offer deep insights to rationally guide further optimization in CE.

In this talk, I will introduce our recent advances in systematically revealing the underlying chemistry and formation of FSI-derived SEI at the interface for Li metal anode in contact with various liquid electrolytes. Based on Cu current collector, we introduce a non-washing protocol of XPS analysis to capture and identify key reaction intermediates from SEI chemistry, allowing precise determination of decomposition pathway for $FSI^-$ anion. Importantly, we reveal a highly dynamic picture at electrode/electrolyte interface where these reaction products can undergo significant dissolution into liquid layer, whereas their relative solubility determines their contribution towards electrode surface passivation. Further, we demonstrate higher-performing electrolytes exhibit faster interfacial passivation that strongly correlates with the reduced solubility of various inorganic species from electrode surface to liquid. These new insights into the SEI chemistry and formation will point out new directions of improving the LMA performance by minimizing side reactions at interface.

## 212g Interface Engineering in Low-PGM Catalysts for Direct Ethanol Fuel Cells

*Yang Yang*

A new strategy to controllably regulate the local bonding states and local coordination environment (LCE) in the classic carbon-supported metal catalysts for direct ethanol fuel cells will be discussed. Particularly, we use F-doped Pd-X-C (X = N, P, S, B) as model systems to study the LCE effects on oxygen reduction reaction (ORR) and ethanol oxidation reaction (EOR). The N-rich Pd surface with regulated LCE not only boosts the activity of Pd but also enhances the catalyst stability by inhibiting the migration and agglomeration of Pd as well as endowing carbon support with long-term anti-corrosion properties. As a result, an unbeatable power density of 0.57 W $cm^{-2}$ with more than 5,900 hours of operation was achieved in direct ethanol fuel cells using the proposed LCE-regulated catalysts, remarkably outperforming benchmarking catalysts. In an extended study, the concept of F-induced LCE regulation can be applied to other Pd-X-C catalysts (X = P, S, B) with drastically improved activities and stabilities.

## 212h Ion Transport and Interfacial Dynamics in Polymerized Ionic Liquids

*Javad Jeddi, Joshua Sangoro*

Polymerized ionic liquids are a class of novel functional polymer electrolytes that combines the unique physicochemical properties of molecular ionic liquids (e.g. non-flammability, wide electrochemical windows, negligible vapor pressures, and ionic conduction) with the outstanding mechanical characteristics of polymers. These materials are promising for a variety of applications including dye-sensitized solar cells, portable batteries, actuators, field-effect transistors and electrochromic devices. However, their ionic conductivity, which is one of the most critical properties in the context of electrochemical energy applications, drops by many orders of magnitude in comparison to their low molecular weight counterparts upon polymerization. In this talk, a new approach to employ nanoscale confinement and interfacial forces to develop polymerized ionic liquids with enhanced ionic conduction will be presented. We will demonstrate that polymerization in nanochannels promotes molecular alignment and charge transport, resulting into enhancement of the dc ionic conductivity by up to three orders of magnitude in model systems. These results contribute to the understanding of the dynamics of nanoconfined molecules and show that in situ polymerization under nanoscale geometric confinement is a promising path toward enhancing ion conductivity in polymer electrolytes.

# 212i Tuning Au@Pd Nanoparticles Interfacial Engineering for Plasmon-Enhanced Hydrogen Evolution Performance

*Lemma Tufa*

The properties of core-shell plasmonic nanomaterials, which result in hot charge-carrier generation, near-field enhancement, and injection, and elevation of local temperature, make them important candidates for electrocatalysis [1]. However, it remains unclear how plasmonic cores and shells influence catalysis performance. Herein, we present a plasmon enhanced phocatalysis of hydrogen evolution reaction (HER) controlled by an interface controlled coreshell structure of Au@Pd NPS. The Au@Pd core-shell structures demonstrated effective photoelectrocatalytic activities toward HER, whereas the catalytic activity was optimized by tuning the amount of Pd nanoparticle-coated Au nanoparticle surface. The results showed that the reaction rate decreased as the shell thickness increased. Core-shell structure-induced lattice strain and energetic hot electron are attributed to such intriguing photoelectrocatalytic performances. The results provide a reliable means to design plasmonic core-shell bimetallic nanostructures with excellent catalytic activity and durability.

# 126b Modeling of Spatiotemporal Temperature Distribution in Hybrid Nanoscale Multifunctional Material for Desorption of $CO_2$

*Raj Panthesh Bhavsar, Bhavana Bhadriraju, Ga Hyun Lee, Ah-Hyung Park, Joseph Kwon*

The atmospheric $CO_2$ ppm levels have been rising alarmingly every year, increasing from less than 300 ppm in 1950 to a record 414 ppm in 2022. To prevent climate change and achieve the net-zero target by 2050, $CO_2$ emissions must be kept in check. This necessitates the development of highly-engineered solutions in the form of Negative Emission Technologies (NET) [1-3]. NETs mainly focus on capturing carbon dioxide either from point source (industrial emissions) or directly from the air. Among the available NETs, Direct Air Capture (DAC) has gained significant attention as it captures low concentrations of $CO_2$ directly from the atmosphere using chemical adsorbents [4]. Specifically, special hybrid nanoscale multifunctional materials are used as adsorbents in DAC to capture $CO_2$ because of their enhanced binding energy, mass transfer, and reaction kinetics [5]. However, regeneration of these materials requires a large amount of energy in the form of conventional heating for $CO_2$ desorption. As an alternative, microwave heating has been employed for adsorbent regeneration to reduce processing time and energy intensity of $CO_2$ desorption process. However, microwave heating poses a challenge of material degradation when exposed to repeated desorption cycles at high temperatures. Therefore, it becomes crucial to understand the temperature variation within the adsorbent.

Motivated by this objective, we model the spatiotemporal distribution of temperature inside a hybrid nanoscale multifunctional material during microwave heating. The material considered in this study is a Solvent Impregnated Polymer (SIP), which encapsulates Polyethylenimine with Nanoparticle Organic Hybrid Material (NOHM) [6]. A ferromagnetic additive is combined with the SIP to enhance targeted heating within the adsorbent. The temperature of the SIP placed inside the microwave is a critical parameter for optimally reducing material degradation. Thus, performing the spatiotemporal temperature analysis will help minimize material degradation and reduce material costs. During microwave heating, electromagnetic waves generate electric and magnetic fields [7-8]. These fields cause heat generation inside the adsorbent placed in the microwave, subsequently releasing the adsorbed $CO_2$ through the principle of targeted heating. The targeted heating at adsorbed sites (NOHM-activated) is caused by the presence of ferromagnetic additive ($Fe_3O_4$ nanoparticles), which results in hot spots inside the adsorbent. In our model, we solve heat diffusion and Maxwell's equations to analyze the temperature variation inside the SIP system during targeted heating occurs. Specifically, we use high-resolution modeling to estimate the spatiotemporal temperature distribution in a magnified polymer system with $Fe_3O_4$ nanoparticle at its core. This approach allows us to identify multiple hot spot regions inside the adsorbent and optimally control the temperature, enhancing the microwave heating process. The effectiveness of the developed model is verified by comparing the hot spots identified by the model with those detected in experiments. Further, the experimental proceedings will incorporate the temperature findings from the modeling to control the overall temperature during desorption in accordance with the regeneration temperature, thus reducing material degradation.

**Literature cited:**

1. Jackson, R.B., Le Quéré, C., Andrew, R.M., Canadell, J.G., Peters, G.P., Roy, J. and Wu, L., 2017. Warning signs for stabilizing global $CO_2$ Environmental Research Letters, 12(11),

p.110202.
2. Board, O.S. and National Academies of Sciences, Engineering, and Medicine, 2019. Negative emissions technologies and reliable sequestration: A research agenda.
3. Keith, D.W., 2009. Why capture $CO_2$ from the atmosphere?. Science, 325(5948), pp.1654-1655.
4. Wu, J., Zhu, X., Chen, Y., Wang, R. and Ge, T., 2022. The analysis and evaluation of direct air capture adsorbents on the material characterization level. Chemical Engineering Journal, 450, p.137958.
5. Rim, G., Feric, T.G., Moore, T. and Park, A.H.A., 2021. Solvent impregnated polymers loaded with liquid-like nanoparticle organic hybrid materials for enhanced kinetics of direct air capture and point source $CO_2$ Advanced Functional Materials, 31(21), p.2010047.
6. Moore, T., Rim, G., Park, A.H.A., Mumford, K.A., Stevens, G.W. and Webley, P.A., 2022. Encapsulation of highly viscous $CO_2$ capture solvents for enhanced capture kinetics: Modeling investigation of mass transfer mechanisms. Chemical Engineering Journal, 428, p.131603.
7. Kumar, D., Dhar, P. and Paul, A., 2021. Thermal Response of Dielectric Nanoparticle-Infused Tissue Phantoms During Microwave-Assisted Hyperthermia. Journal of Thermal Science and Engineering Applications, 13(6).
8. Xiao, Y. and Shao, Z., 2023. Numerical analysis of mortar-aggregate separation induced by microwave heating. International Journal of Thermal Sciences, 184, p.107957.

# 126c Structure and Transport Behaviors of Liquid-like Nanoparticle Organic Hybrid Materials Designed for Combined $CO_2$ Capture and Conversion

*Ga Hyun Lee, Tony Feric, Sara Hamilton, Xiaoyang Shi, Ah-Hyung Park*

Liquid-like Nanoscale Organic Hybrid Materials (NOHMs) consisting of polymer grafted nanoparticles have shown great promise in applications, such as electrochemistry and gas separation due to their enhanced conductivity, tunability, and negligible vapor pressure. Recently, NOHMs are considered to be used as novel $CO_2$ capture materials and electrolytes for $CO_2$ conversion to high value chemicals. However, to employ NOHMs as $CO_2$ capture materials and electrolytes it is important to understand the conformation and transport behaviors of NOHMs in different chemical milieu. Our studies indicate that in the aqueous environment, there exists a large amount of free polymer in the solution that is not grafted to the functionalized nanoparticles. These protonated free polymers, dispersed in the aqueous solvent, may also strongly interact with the grafted polymer layer and greatly affect the neat structure of NOHMs. Thus, we identified those as "interacting" polymers to distinguish them from tethered or truly free polymers in the fluid system. The presence of supporting electrolyte shows a greater effect on the structure of NOHMs-based fluid as it not only alters the structure of the free polymer but also hinders the interaction of the polymer with the functionalized nanoparticles. Moreover, the change in the interaction of the Jeffamine M2070 with the functionalized nanoparticles due to the addition of supporting electrolyte has revealed a drastic change in the viscosity of NOHMs solutions. Overall, the dispersion of the free polymer, the interaction of the interacting polymer with grafted polymer as well as the change in conformation of free polymer and grafted layers with the addition of different solvents and salts provides valuable insight into how these nanoscale hybrid materials interact with $CO_2$. These results can be applied to fine-tune the structure of liquid-like NOHMs and will aid in a better understanding of their performance as $CO_2$ capture materials and electrolyte.

## 126e Ultrathin: Understanding the Nanostructure of Alumina Atomic Layer Deposited Films on Layered Cathode Oxides

*Julie A. Nguyen, Elizabeth K. Allan-Cole, Meghan N. Herbert-Walters, Madison B. Martinez, Barbara K. Hughes, Michael F. Toney, Hendrik Heinz, Alan Weimer*

Atomic layer deposition (ALD) is a surface deposition technique that has been shown on layered transition metal oxides, such as $LiCoO2$ and $LiNixMnyCo1-x-yO2$, to improve the cycling stability and reduce the capacity fade of the resulting Li-ion batteries. In particular, thin (< 2 nm thick) alumina ALD films have been experimentally shown to be beneficial, with evidence that a Li-Al-O film is formed. Since the cathode surface forms an interface with electrolyte in a battery, it is important for this surface to promote a stable interface. ALD modifies the transition metal oxide surface. Our present work investigates alumina ALD on layered transition metal oxides and seeks to address why a Li-Al-O film is formed and how it is beneficial to the cathode oxide. In this work, we employed molecular dynamics (MD) simulations to explore the nanostructure of a Li-Al-O film in contact with a layered cathode oxide.

The length scale of MD simulation is on the order of nanometers, which matches the length scale of the ALD films investigated. The Li-Al-O film in contact with a layered cathode oxide was simulated by bringing an alumina surface (1-10 monolayers thick) into contact with a LiCoO2 surface and performing a sequence of annealing steps in the NVT ensemble (where the number of moles, the volume, and temperature are held constant) to generate disordered Li-Al-O films. System properties such as film density, energy of Li inclusion, element mass densities in the z direction, and atom coordination number were obtained from subsequent NVT production runs. The Interface Force Field in the 12-6 Lennard Jones form was used to model both the alumina and the lithium cobalt oxide.

The simulated film density agreed with the reported experimental density of alumina ALD films. Element mass density profiles in the z direction revealed layering of Al and O close to the LiCoO2 surface. This layering eventually transitioned to an isotropic disordered structure that agreed with conventional understanding that ALD alumina films are amorphous. Aluminum atoms with coordination number of four were the predominant species in the film bulk.

Our present work provides atomic insight into the experimentally observed Li-Al-O film. This work helps inform the understanding o f the protective mechanism of alumina ALD films on layered transition metal oxides for Li-ion battery applications.

## 126f Molecular Design of Electrochemical Interfaces for Sustainable Separations in Clean Energy

*Xiao Su*

Molecular design of electrochemical interfaces is critical for the development of new separation processes for the clean energy transition. Through electrochemical control, selective adsorption can be carried out without the need for any chemical eluents or physical processing, leading to modular and sustainable separations. However, significant challenges remain primarily in achieving desired molecular selectivity, especially for minority species in the presence of excess salts. Electroactive interfaces offer an attractive platform to control voltage while dictating the desired ion-selectivity.

Here, we present the tuning of redox-active interfaces for important separations related to clean energy, including critical element recovery and ammonium production. First, we discuss the structural tuning of redox-polymers for achieving finer multicomponent separations. We discuss applications ranging from the electrosorption-based recovery of critical elements to selective electrodeposition for battery recycling. Next, we discuss an electrified approach for integrating reaction and separations, especially the capture of nitrate from waste runoffs and the tandem upcycling to value-added ammonium. We design a bifunctional redox-electrode in which the electrosorbent preferentially binds nitrate during oxidation, and a co-localized electrocatalyst reduces the nitrate to ammonium during electrochemical release. We elucidate the potential and pH-responsive of redox-films for the binding of nitrate, and the underlying molecular mechanisms. We demonstrate the capability of a single electrochemical cell to greatly enhance the energy efficiency towards nitrate capture and ammonium conversion.

Overall, we highlight the promising role of electroactive interfaces for enabling energy-relevant separations and catalytic transformations.

# 126g Novel Electrocatalysts for Nitrogen Oxides Reduction at Ambient Conditions

*Feng Jiao*

Mitigating nitrogen oxides ($NO_x$) emissions is critical to tackle global warming and improve air quality.[1]Conventional $NO_x$ abatement technologies for emission control suffer from a low efficiency at near ambient temperatures.[2] Electrochemical $NO_x$ reduction reaction is an alternative approach to convert $NO_x$ into benign $N_2$ or useful chemicals (e.g., $NH_3$), powered by renewable electricity. For environmental applications, $N_2$ is the desired product because it is non-toxic and can be emitted directly to the atmosphere, whereas $NH_3$ is preferable when local generation of fertilizer is needed for crops. More importantly, Electrochemical $NO_x$ reduction reaction can be conducted at high reaction rates at ambient temperature because it is driven by applied potential (i.e., electrical energy), as opposed to high temperatures and pressures in conventional. In this work, we investigate a series of transition metal catalysts for electrocatalytic reduction of gaseous nitric oxide (NO) and nitrous oxide ($N_2O$) at ambient condition in a gas-fed electrolyzer configuration.[3] Fe, Co, Ni, Cu, Pd, Ag, and Pt were chosen as representative metal catalysts because they are common catalysts for electroreduction reactions, such as oxygen reduction reaction and hydrogen evolution reaction (HER). To gain mechanistic insights, NO partial pressure and pH effects are investigated on Pd and Cu, representing catalysts selective for $N_2$ and $NH_3$ formation, respectively. Furthermore, combining various electrochemical experiments and *operando* flow electrolyzer mass spectrometry (FEMS), we determined reaction pathways and rate determining steps (RDSs) of NO electroreduction on Pd and Cu.

References

1. Gholami, F., Tomas, M., Gholami, Z. & Vakili, M. Technologies for the nitrogen oxides reduction from flue gas: A review. *Total Environ.* **714**, 136712 (2020).
2. Yadav, D. & Prasad, R. Low temperature de-$NO_x$ technology-a challenge for vehicular exhaust and its remedation: An overview. *Procedia Technol.* **24**, 639–644 (2016).
3. Ko, B. H., Hasa, B., Shin, H., Zhao, Y. R. & Jiao, F.* Electrochemical reduction of gaseous nitrogen oxides on transition metals at ambient conditions. *Amer. Chem. Soc.* 144, 1258-1266 (2022).

## 24a Chemical Processing of Function Accelerated Nano-Materials at the Atomic Scale

*Jane Chang*

This talk presents current advances in atomic scale processing of function accelerated nano-materials and the key challenges and opportunities in the future that can enable novel integrated circuits for commercial and defense applications. Amongst all enablers, plasma processing science, due to its intrinsic nonequilibrium nature, has been at the heart of many key technologies and provided breakthrough solutions in the information processing system hierarchy, translating physics to a physically functioning embodiment with enhanced device performance. To realize the full potential of these materials, plasma surface science becomes increasing more important in material synthesis, material processing and device embodiment, all requiring atomic level control and precision. In this talk, plasma enhanced atomic layer deposition of multiferroic materials and atomic layer etching of magnetic materials will be showcased as examples that leverage the fundamental knowledge base in plasma physics and chemistry to address future technological grand challenges.

## 24b Surface Reaction Kinetics during Area-Selective Thin Film Deposition

*Gregory N. Parsons*

Advances in chemical reaction engineering impact chemical manufacturing, but also play an important role in manufacturing of advanced semiconductor devices. Chemical processes used for semiconductors, including lithographic patterning, thin film atomic layer deposition (ALD) and atomic layer etching (ALE), are among the most precise chemical methods used in any large-scale manufacturing enterprise. Even so, new problems are arising that demand better control, precision and flexibility of deposition and etching. A key challenge is the ability to achieve desired atomic structure and bonding while maintaining low process temperature, often less than 200°C. Compared to high temperature processes where thermodynamic equilibrium can be attained, controlling reactions at low temperature generally requires better control and adjustment of surface reaction kinetics.

This presentation will discuss the role of reaction kinetic mechanisms in controlling thin film nucleation and area-selective deposition (ASD) for applications in electronic materials nanoscale patterning. For example, reaction rates during ALD can be independent of temperature, whereas the extent of material removed during ALE is generally temperature dependent. ASD often proceeds by combining ALD and ALE reactions, so understanding these
rate processes is instrumental in process design. Moreover, the balance between deposition and etching rates can lead to some unexpected results when deposition and etching are applied as an "intensified" process, either in parallel or sequentially, to patterned substrates. John Ekerdt's substantial contributions to reaction rate mechanisms for electronic materials have influenced our group's thinking and understanding, and we are happy to participate in this special session in his honor.

## 24c Controlling Nucleation and Selectivity in Atomic Layer Deposition for Materials Synthesis

*Stacey Bent*

With the growing demands on materials synthesis for applications such as advanced electronics and energy, atomic layer deposition (ALD) continues to gain attention as an important method to achieve nanoscale features at the sub-10 nm length scale. Based on sequential, self-limiting vapor-surface reactions, ALD offers exceptional conformality, thickness control at the Angstrom level, and tunable film composition. This talk will describe research into fundamental interfacial reaction mechanisms that underlie ALD. A deep understanding of the vapor-surface chemistry behind ALD can be critical toward developing both improved ALD nucleation and the associated process of area selective atomic layer deposition (AS-ALD). This talk will provide examples of tuning surface chemistry to either enhance or inhibit ALD nucleation. Some surface functionalization can lead to enhanced nucleation and better film quality. On the other hand, the development of inhibitory layers can alter the native surface reactivity to block nucleation. This inhibition strategy enables good selectivity in AS-ALD of thin films on a variety of substrate materials, including dielectrics and metals.

## 24d High Quality Tunable Plasmonic Nanostructures By Atomic Layer Deposition

*Brian Willis, John A. Grasso, Rahul Raman*

Nanostructures made of Cu, Ag, and Au exhibit localized plasmon resonances that are tunable through visible and infrared wavelengths. Plasmonic effects enhance the interaction of light and matter, and nanostructures can act as tiny antenna to collect light and focus it at the nanoscale. The broad spectral tuning range and overlap with the solar spectrum have stimulated applications in energy harvesting, photocatalysis, photodetection, spectroscopy, and novel molecular scale devices. Many of these applications benefit from intense local electric fields generated by plasmons at resonance. These enhanced fields influence molecular transitions and may generate highly energetic carriers as well as localized heating. Field enhancements are most intense in nanoscale junctions between neighboring particles, and there has been much effort to fabricate nanoscale junctions between plasmonic particles. Modern nanofabrication methods are generally limited to minimum feature sizes of 10 nm, which makes it difficult to achieve the highest field enhancements that occur for smaller nanojunctions. Atomic layer deposition (ALD) is a proven tool for highly precise thin film growth rates and thickness control, and it can be applied to arbitrary sizes and shapes of nanostructures, including lithographically fabricated nanostructures for nanodevices. In this paper, we present work on area selective ALD of Cu metal to create new types of plasmonic nanostructures, including nanostructures with electrical interconnects. We investigate several strategies to integrate Cu ALD with Pd and Au materials for achieving nanoscale junctions with plasmonic properties, and compare experimental measurements with finite difference time domain simulations. We demonstrate how ALD can achieve sub-10 nm nanojunctions and study their optical properties and plasmonic characteristics. We demonstrate single nanometer size nanojunctions with high quality plasmonic resonances. Our studies indicate that thermal effects and alloying strongly affect plasmonic properties, and we identify areas for improvement. The findings are promising for new devices that combine plasmon resonances with electrical functions.

## 24e Chemical "Engineerifying" Electronics

*Michael A. Filler*

This talk will overview our recent efforts to chemical "engineerify" electronics manufacturing. While chemical engineering has played a major role in the development and success of integrated circuitry, the process by which we manufacture integrated circuits – the planar process – is strikingly antithetical to the foundational tenets of chemical engineering. To this end, we are developing a suite of processes to modularize nanoelectronic devices and hyper-scale their manufacturing. Modularization at the device level promises an unprecedented combination of performance, cost, and function; opening the door to electronic systems that are natively heterogeneous, on-demand manufacturable, physically reconfigurable, and/or dispersible. This "nanomodular" device fabrication synergistically combines multiple bottom-up and area-selective processes: (i) vapor-liquid-solid semiconductor nanowire growth, (ii) a new nanoscale polymer masking process, and (iii) area-selective atomic layer deposition. Our approach yields single-crystalline, high mobility nanowires with nanoscale coaxial thin films self-aligned to the internal dopant profile. In parallel, we are developing the Geode process to increase manufacturing throughput by orders-of-magnitude. This scale-up is made possible by an unconventional substrate – the interior surface of hollow silica microcapsule powders – on which nanowire growth and subsequent processing occurs. Collaborative efforts are also enabling nanowire property characterization in a high-throughput, non-contact fashion and high-resolution interconnection of nanomodular devices to form functional circuitry.

## 24f The Elusive Surface M2O7 Dimer Active Site

*Israel Wachs*

The presence of dimeric surface $M_2O_7$ ($O_3M$-O-$MO_3$) active sites have been proposed over the years to be present in numerous supported metal oxide catalyst systems that find application in energy solutions (e.g., conversion of $CH_4$ to $C_2H_4$/$C_2H_6$, $CH_4$ to aromatics, $CH_3OH$ oxidation, selective catalytic reduction of $NO_x$, etc.). With advances in modern molecular spectroscopic approaches, it is now possible to examine the molecular and electronic structure of the supported metal oxide active sites under controlled environmental conditions. This presentation will examine the nature of the surface metal oxide active sites for several systems where dimeric surface $M_2O_7$ active sites have been proposed.

## 24g The Dual Nature of Chemical Reaction Engineering

*James Rawlings*

A textbook by our honoree starts with the following statement: 'Chemical reactors are at the core of the chemical engineering discipline, and chemical reactor analysis and design is one of the distinguishing courses that clearly separates the chemical engineers from the other engineering professionals.'

This talk surveys the experiences of the honoree and one of his co-authors in creating a textbook for this subject. We present many of the behind-the-scenes challenges that were faced and pedagogical choices that were made during that book writing journey.

Next some discussion of joint research on the topic of chemical vapor deposition of silicon and germanium thin films is presented.

From these teaching and research perspectives some lessons on the dual nature of the subject of chemical reaction engineering are revealed.

## 24h Plasma Enhanced Atomic Layer Deposition of Silicon Carbonitride

*Samuel Johnson, Tsung-Hsuan Yang, Jianping Zhao, Toshihiko Iwao, Charles Schlechte, John Carroll, Gabriel Blankemeyer, Peter Ventzek, Joaquin Resasco, Gyeong S. Hwang, John Ekerdt*

Conformal deposition of dielectrics is necessary for many applications, including as spacers for self-aligned multiple patterning and as charge trap layers in NAND memory features. Plasma enhanced atomic layer deposition (PEALD) of silicon nitride is promising for these applications, satisfying many requirements of deposition. However, silicon nitride films deposited with this technique typically have low hydrofluoric acid wet etch resistance, limiting its applications. Incorporating carbon into silicon nitride to form silicon carbonitride (SiCN) films can overcome this shortcoming by increasing wet etch resistance, but it also increases leakage currents. By incorporating small amounts of carbon to form silicon carbonitride films, a large increase in etch resistance can be achieved with only a small increase in leakage current. Finding a process that delivers a ternary component film with specific material properties is inherently complex as both stoichiometry and bonding nature need to be controlled through surface reactions. Managing the carbon content and nature of bonding within SiCN films are the focus of this talk.

We use PEALD to deposit silicon carbonitride by alternating a thermal exposure of a silane derived single source precursor, bis(dimethylamino)dimethylsilane, containing Si-N, Si-C, and N-C bonds at 100 °C and a plasma step as a model system for SiCN deposition. A plasma step is needed to create a surface on which the precursor can chemisorb leading to the desired film composition. An experimental capacitively coupled radio frequency plasma source is used to understand how different plasmas, including ammonia, hydrogen, nitrogen, and pure argon impact the nature of chemisorption and film nature. Films were characterized using *in situ* Fourier transform infrared spectroscopy (FTIR) and *in situ* X-ray photoelectron spectroscopy. First principles simulations are used to assess the fundamental mechanisms at play.

Changing the plasma condition had a large effect on carbon concentrations of deposited films, changing the C:Si ratio from 0 with an $NH_3$/Ar plasma to 4.3 with a $N_2$/Ar plasma. In addition to converting precursor-based adsorbates into a film, the choice of plasma changes the surface modalities ($NH_x$, undercoordinated Si-N, or Si-H modes), thereby altering the precursor-surface interactions during the adsorption step. This principle was exploited to tune film compositions to realize carbon concentrations ranging from C:Si ratios between 0.1 to 0.7 by using multiple plasma exposures to generate mixed modality surfaces for precursor adsorption.

# AUTHOR INDEX


Abbas, Zain....7
Aglave, Ravindra....9
Akintola, Ibukunoluwa....3
Alcantar, Norma....8
Al-Kaysi, Rabih O....16
Allan-Cole, Elizabeth K....27
Alrashdi, Khuloud....8
Ayub, Khurram Shahzad....7
Bandi, Bharathi....18
Bao, Zhenan....20
Barboun, Patrick....6
Bardeen, Christopher J....16
Bayer, Brian....2
Bent, Stacey....32
Bhadriraju, Bhavana....24
Bhan, Aditya....2
Bhavsar, Raj Panthesh....24
Blankemeyer, Gabriel....37
Boioli, Francesca....9
Bragg, Arthur....12
Bruggeman, Peter....2
Carroll, John....37
Chang, Jane....30
Chatterjee, Abhijit....18
Che, Fanglin....4
Chen, Han-Ting....5
Choi, Joshua....12
Conley, Ashley....12
Correa-Baena, Juan-Pablo....11
Cui, Yi....20
Dziatko, Rachel....12
Ekerdt, John....37
Feric, Tony....26
Filler, Michael A....34
Ghate, Pranaya....16
Giri, Gaurav....12, 15
Go, David....1, 3
Goswami, Anshuman....6
Grasso, John A....33
Greeley, Jeffrey....19
Guthrie, Stephanie....12
Hajian, Moein Z.D....14
Hamilton, Sara....26
Heinz, Hendrik....27
Herbert-Walters, Meghan N....27
Hicks, Jason....3, 6
Howe, Joshua....14
Hughes, Barbara K....27
Hussain, Syed....17
Hwang, Gyeong S....37
Ibrahim, Saleh Ahmat....4
Iwao, Toshihiko....37
Jeddi, Javad....22
Jiao, Feng....29
Johnson, Samuel....37
Kamal, Muhammad Shahzad....17
Kim, Sanggyun....11
Kwon, Joseph....24
Lee, Dongho....5
Lee, Ga Hyun....24, 26
Lee, Garam....1
Lee, Hyunhee....14
Linic, Suljo....5
Ma, Hanyu....6
Mallick, Shikha....9
Martinez, Madison B....27
Mas, Peter....9
Maxson, Tristan....13
Nguyen, Julie A....27
Nikolla, Eranda....19
O'Brien, Casey....1
Oh, Donghyun....11
Otor, Hope....6
Paolucci, Christopher....12
Park, Ah-Hyung....24, 26
Parsons, Gregory N....31
Patil, Shirish....17
Perini, Carlo Andrea Riccardo....11
Petris, Panos....9
Phister, Meagan....12, 15
Raman, Rahul....33
Rawlings, James....36
Resasco, Joaquin....37
Rivera-Castro, Gerardo....3
Sahinidis, Nikolaos....11
Sangoro, Joshua....22
Schlechte, Charles....37
Schneider, William....6
Shi, Xiaoyang....26
Smith, Natalie....15
Smith, Zachary P....14
Stone, Kevin....12
Su, Xiao....28
Sviripa, Anna....12
Szilvasi, Tibor....13
Toney, Michael F....27

Trelles, Juan Pablo ........ 4
Tufa, Lemma ........ 23
Velinkar, Kunal ........ 19
Ventzek, Peter ........ 37
Von Gunten, Alex ........ 19
Wachs, Israel ........ 35
Weimer, Alan ........ 27
Westendorff, Karl ........ 12
Willis, Brian ........ 33
Wong, Hsi-Wu ........ 4
Xiao, Deng ........ 17
Yang, Tsung-Hsuan ........ 37
Yang, Yang ........ 21
Yu, Weilai ........ 20
Zaman, Waqas Qamar ........ 7
Zhao, Jianping ........ 37